Renewable Energy Storage

Based on papers presented at the one-day Seminar *Energy Storage – Its Role in Renewables and Future Electricity Markets*, held at the IMechE Headquarters, London, UK, on 15 December 1999.

IMechE
Seminar Publication

I MECH E

Renewable Energy Storage
Its Role in Renewables and Future
Electricity Markets

Organized by
The Research and Technology Committee of
The Institution of Mechanical Engineers (IMechE)

Supported by
The Engineering and Physical Sciences Research Council (EPSRC)
The Office of Science and Technology's Foresight Initiative

IMechE Seminar Publication 2000–7

**Professional
Engineering
Publishing**

Published by Professional Engineering Publishing Limited for The Institution of
Mechanical Engineers, Bury St Edmunds and London, UK.

First Published 2000

ISSN 1357–9193
ISBN 1 86058 306 7

A CIP catalogue record for this book is available from the British Library.

Related Titles of Interest

Title	Editor/Author	ISBN
Wind Energy 1999 – Power from Wind	Peter Hinson	1 86058 206 0
Handbook of Mechanical Works Inspection – A guide to effective Practice	Clifford Matthews	1 86058 047 5
IMechE Engineers' Data Book	Clifford Matthews	1 86058 175 7
CHP 2000 – Co-Generation for the 21st Century	IMechE Conference	1 86058 141 2
Energy Saving in the Design and Operation of Compressors	IMechE Seminar	0 85298 985 7
Energy Saving in the Design and Operation of Pumps	IMechE Seminar	1 86058 030 0
Hydro Power Developments – Current Projects, Rehabilitation, and Power Recovery	IMechE Seminar	1 86058 121 8
Investment in Renewable Energy	IMechE Seminar	1 86058 163 3

For the full range of titles published by Professional Engineering Publishing contact:

Sales Department
Professional Engineering Publishing Limited
Northgate Avenue
Bury St Edmunds
Suffolk
IP32 6BW
UK

Tel: +44 (0)1284 724384
Fax: +44 (0)1284 718692

Contents

S684/001/99

Increasing the value of renewable sources with energy storage

N JENKINS and **G STRBAC**
Manchester Centre for Electrical Energy, UMIST, UK

SYNOPSIS

Energy storage has the potential to increase the value of intermittent renewable energy generation in large electric power systems. The characteristics of some of the main renewable sources are reviewed and the opportunities for the use of energy storage in association with renewable generation in a de-regulated power system are discussed.

1   INTRODUCTION

The UK Government is working towards a target of renewable energy providing 10% of UK electricity supplies by 2010, which corresponds to approximately 40 TWh, with an intermediate goal of 5% by 2003 [1]. As the load factor of the majority of new renewable sources is relatively low (say a mean value of 50%), this target of 10 % electrical energy is likely to require the installation of some 9 GW of renewable generating capacity. This capacity is a significant fraction of the summer minimum loading of the UK system, which is under 20 GW. Under extreme circumstances, when low demand for electricity coincides with high output of renewable sources, system stability and frequency control will become of concern and there will be cost implications of such operation due to the increased amount of spinning reserve which will be required. Even without renewable generation the need for flexible generation plant at times of low demand, when nuclear plant and some base load gas stations with limited flexibility represent a high proportion of the generation, has already been recognised [2,3].

Intermittent renewable energy sources are presently connected into the UK power system without significant difficulty and with no requirement for energy storage. Further, at the levels of penetration presently anticipated in the UK, the intermittent nature of renewable energy generation is a commercial/economic question rather than one of fundamental technical limitation. The House of Lords Select Committee report [4] was quite clear that "There are no insuperable problems in operating the UK electricity network with substantial amounts of renewable energy, including intermittent sources, well beyond the present 10% target". This view was shared by National Grid Company [3], who referred in their evidence to the typical winter week day load pick up of 12,000 MW (25% of peak demand) over a two hour period and estimated the cost of making more flexible plant available at "approximately £10 million for an additional 100 MW available all year", under current market conditions. However the ability of the operator (in the UK the National Grid Company) to control the power system effectively is entirely dependent on the availability of suitable plant. At present such plant is available but, if flexible, mid-merit coal fired generation is retired, then there will be a commercial opportunity for other forms of flexible generation, including energy storage equipment.

It is interesting to note the projections in Denmark (Figure 1) where it is anticipated that wind power (onshore and offshore) as well as rural, small-scale, CHP (which is not subject to central despatch) will exceed the system minimum load for considerable periods. Denmark is, perhaps, a rather special case being connected to Germany by a 400 kV AC system and to Norway and Sweden by HVDC circuits but the anticipated level of non-despatched embedded generation is remarkable.

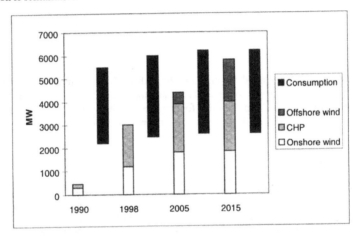

Fig 1 Dispersed generation and consumption in Denmark (after [5] and personal communication, Mr A Sorensen, ELTRA)

Renewable energy generation has an important role in assisting the UK in meeting its commitments in reducing emissions of greenhouse gases [1,6]. Although renewable generation reduces production of *energy* from conventional sources, intermittent sources, (i.e. wind, photovoltaics, wave and run-of-the-river hydro) may not capable of displacing significant generating plant *capacity* due to their inherent intermittent nature and therefore, limited ability to provide a continuous supply and support system security requirements. Even in an ultimate renewable energy based system (with renewable generation producing the vast majority of electricity), considerable capacity of conventional plant may still be required. This would mean the power system acting as a backup or standby system, which obviously reduces the overall value of renewable generation. Clearly, with high penetrations of renewable

    S684/001/99   © IMechE 2000

energy the economic benefits of sources such as wind, solar and wave for electrical power generation may be considerably limited by the volatile nature of the availability of the primary supply.

Energy storage can address these problems. Many technologies have been developed with the aim of offering storage facilities. However, only pumped hydro schemes have achieved real penetration and acceptance due to their ability to provide both large power and energy outputs, although this technology is fundamentally constrained by geography and cost. Alternatives are now emerging with similar capabilities but without the siting and environmental constraints and with potentially lower cost. Clearly, such technology offers the possibility to enhance the value of renewable generation and enable it to contribute a more predictable capacity to the system. Furthermore, due to its rapid response and fast ramp rates, effective large scale energy storage offers a range of associated benefits, particularly in the area of system frequency control, reserve requirements and network reinforcement in case of distributed storage schemes.

2   CHARACTERISTICS OF RENEWABLE ENERGY SOURCES

The White Paper issued by the European Commission in 1997 [7] attempted to predict which renewable generation technologies could make a major contribution to European energy supplies. It indicated that biomass and wind energy were the two technologies likely to make the largest increase in contribution to renewable energy generation. Solar photovoltaics was also shown to show a very large increase but from a low base.

Table 1 Estimated contribution to European Union electrical energy supply [4, 7]

Energy Source	1995 TWh	2010 TWh
Wind	4	80
Hydro	307	355
- Large (10 MW plus)	270	300
- Small (under 10 MW)	37	55
Solar photovoltaic	0.03	3
Biomass	22.5	230
Geothermal	3.5	7

Biomass generation is, of course dependent on its energy source but is generally less intermittent than wind or solar photovoltaics. Depending on the specific fuel source it can be considered to have a degree of storage included [8]. However, due its relatively low energy density, storage of large quantities of biomass is difficult and expensive.

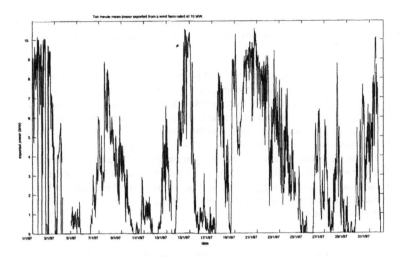

**Figure 2 Time series output of a UK wind farm over a month
(courtesy of National Wind Power)**

Figure 2 shows the output of a 10 MW wind farm over a winter month. It will be seen that rated output is reached on a number of days but there are also times of no generation. Figure 3 shows the output of a small (1.8 kWp) test array of photovoltaic generator. There is a clear daily variation in output but also the effect of cloud may be seen on the 6th day.

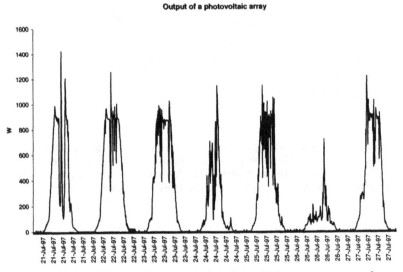

**Figure 3 Time series output of a photovoltaic array over a week
(courtesy of Dr T Markvart, Southampton University)**

S684/001/99

3   OPPORTUNITIES FOR ENERGY STORAGE WITH RENEWABLES

Intermittent renewable energy generation may be characterised by: (1) high capital cost and so the requirement to operate the plant whenever the resource is available, (2) remote locations often away from strong electrical infrastructure (3) sometimes poor power quality (4) intermittent output and (5) low load factors.

The opportunities for energy storage to enhance the value of electric power systems have recently received very considerable attention [9-12]. Some of these may be listed as [11]:
- Operating reserve, against the loss of the largest generator
- Load following and maintaining power flows across critical interconnections
- Black start to re-energise critical generating station loads
- Voltage control, particularly on resistive networks
- Loss reduction, by reducing peak flows
- Investment deferrals
        generating plant
        transmission and distribution circuit capacity
- Power Quality improvement
- Reliability enhancement

However, following the paradigm shift in which electrical energy as a product is separated from its delivery as a service it is necessary to re-classify the contribution which energy storage can make to renewable generation. The fundamental division is, of course, between Energy Trading and Transport Services. In the UK the commercial arrangements for trading electrical energy are presently undergoing a major change with the Pool mechanism being replaced in November 2000 by a system based on Bilateral Contracts, a Power Exchange and a Short Term Balancing Market. This new trading system, referred to as the New Electricity Trading Arrangements (NETA) is still under development and its impact is not yet known. However, it is likely to reward predictability and controllability in generation as it is necessary for energy suppliers to balance generation and demand in every half hour period. The commercial consequences for intermittent renewable energy generators has yet to be determined but some estimates are that up to 25% of the price paid for wind generated energy will be lost due to penalties incurred in the NETA Balancing Market. This represents an important potential opportunity for energy storage. Similarly, energy storage can be used for the management of constraints on the transmission and distribution networks. In the light of the proposed auction mechanism for access to the transmission system, in the short term, energy storage could take an active role in the transmission access rights market. This concept is, of course, analogous to the concept of energy arbitrage, but in this case the commodity traded is access to the transmission network.

However, in view of the uncertainty and the continuing evolution of these new commercial arrangements, this paper focuses on the contribution of storage to Transport Services which are further grouped as shown below. The importance of this classification is that it is the conventional commercial structure found in modern de-regulated power systems.

Connection Charges
- Power Quality improvement
- Optimal use of circuit capacity and local load matching
- Voltage and angular stability enhancement

Use of Network Charges
- Distribution Use of System charges
- Transmission Use of System charges
- Network reliability enhancement
- Network losses

Generation Ancillary Services
- Black start
- Generation reserve
- Frequency control

Each of the above are discussed briefly below. The immediate commercial opportunities for energy storage would appear to be with respect to generator connection where energy storage has the possibility to increase the capacity of renewable generating plant that may be connected at a voltage level.

3.1 Connection charges

In UK distribution networks, connection charges are levied on a "deep charging" basis and the generator has to pay all the additional costs of the network associated with the connection. This leads to strong pressure to minimise connection costs and connect the generation at the lowest possible voltage level (and hence at relatively low short-circuit levels).

A number of new renewable generator technologies deliver rapidly varying output power (e.g. wind, wave and photovoltaics). When a large number of relatively small generators are connected to a strong electrical network then induced voltage "flicker" tends not to be a constraint on the connection of the generators as other voltage limits (e.g. voltage rise or stability) tend to be reached first. However for large single installations such as a single large turbine, large photovoltaic array or wave generator connected to a high source impedance network then voltage flicker is often the limiting constraint. In such distribution networks it is common to find source impedance X/R ratios as low as unity and so reactive power compensators tend to be of limited effect.

For large numbers of relatively small generators (e.g. wind farms), the limit to the capacity which may be connected is often the steady state voltage rise. This is generally calculated on the basis of minimum network load and maximum generation. Even if the maximum voltage limit is shown to be violated only once a year then permission to connect the generator will be refused. In transmission systems, network voltages may be controlled effectively by adjusting the flow of reactive power. In distribution networks with low X/R ratios this method of voltage control can require very large VAr flows and hence high associated electrical losses.

There are two potential solutions: (1) constraining the generation in response to network voltage conditions and (2) storing energy for release at a later time when the local network load has increased. For occasional constraints it is likely to be more economic to reduce generation but if the constraints occur frequently then energy storage may be justified.

The Danish use of rural CHP systems illustrates an interesting approach to energy storage (even though is it heat energy which is stored) [13]. Medium sized (up to 10 MW) reciprocating engines or gas turbines, fed from natural gas, are used to provide both electrical energy and district heating. Energy storage is used to store heat energy in large hot water

S684/001/99 © IMechE 2000

tanks so that the CHP sets may be run during the times of peak network electrical load and at rated output. Thus, although the energy is stored in the form of heat the storage systems allows the operation of the CHP sets at times of maximum benefit to the electrical distribution system.

A number of renewable energy technologies use induction generators and it has been shown that there is a limit to the capacity which may be connected to the distribution network before voltage instability becomes a potential hazard [14]. This has been identified as a significant issue for the connection of large offshore wind farms using AC circuits. Although some improvement may be obtained using reactive power compensation it is likely that much greater improvements in stability could be achieved using real power injection and energy storage.

At present the question of angular (transient) stability of synchronous dispersed renewable generation is of limited interest as if the generation trips for a remote fault then it is re-synchronised after some while and only some renewable energy is lost. However, once dispersed renewable generation starts to comprise a significant fraction of the system generation its dynamic performance will assume considerable importance as loosing generation unnecessarily will no longer be acceptable. Again, energy storage is likely to have a commercial role as the alternatives are increasing the short-circuit capacity of the network or installing faster network protection. Both of these options are likely to be extremely expensive.

3.2 Use of network charges
Energy storage devices can, in principle, be built anywhere. This flexibility may be utilised to increase the benefit of the plant to the distribution and transmission network capital and operating costs and enhance the network reliability performance.

If the use of network charges are cost-reflective, storage devices sited in favourable locations from the network investment and operating cost perspective, should benefit from negative use of system charges. For example, the value of investment in network reinforcement that can be substituted by favourably located energy storage devices should be reflected through the use of system charges. Furthermore, as the regulation of network businesses is focusing more and more on service quality based revenue recovery, the value of storage devices that can provide alternative supplies should also be reflected through use of system charges. At present, network tariffs are only partially cost reflective, but the trends in their development suggest potential opportunities for energy storage. It is, however, important to stress that only bulk energy storage can qualify for the provision of these services.

In the context of the use of energy storage with renewable generation, it is interesting to examine the potential of storage in the context of the UK transmission network. Due to locational development of generation and demand in England, Wales and Scotland, there is a North to South power flow pattern across the transmission system throughout the year. The existing capability of the transmission network is generally sufficient to carry these power flows although there are some occasional restrictions. Due to large flows the incremental differential in losses between North and South is in the range of 8–12 % [2,3], which is the amount for which the value of generation in North is less than that in the South. Hence, energy storage located in the South should benefit through transmission loss adjustment factors, which are yet to be established.

Another issue that is of interest here is the use of favourably located energy storage devices to support unfavourably located renewable sources. Some of future renewable generation technologies may be expected to locate predominantly in the North as a result of higher primary source energy densities [3]. Hence they have the potential to increase considerably North to South power flows and this may cause a need for system reinforcement and lead to higher losses. Energy storage devices could reduce this indirectly by displacing generation located in the North during the periods of peak flows from North to South. Clearly, storage plant located in the South should benefit from negative transmission-use-of-system charges.

For the application of energy storage at the distribution level, it is important to remember that the ability of distribution networks to transport energy is usually voltage limited. Due to the nature of distribution network, both active and reactive power flows are important for the management of voltage profiles on the distribution network. Hence, the use of energy storage could be optimised in order to control the voltage in local networks and hence postpone reinforcement and reduce losses. Again, if the distribution use of system charges were cost reflective, favourably located storage plant should benefit from negative use of system charges. Similarly, any benefit from reduction of losses should be reflected through the appropriate loss adjustment factors.

The focus in present regulation of distribution networks is moving towards network performance, currently measured through Customer Minutes Lost. In cases when available transmission circuits are not able to supply all connected load and some of it must be shed, the ability of storage to reduce the amount of load that must be shed would clearly be very beneficial. The corresponding pricing mechanisms, through adequate use of system charges or distribution network ancillary services contracts should quantify the value of storage in this context. This area is still a research topic and has yet to be fully investigated.

3.3 Generation ancillary services

At present in the UK, renewable generating plants are not despatched and take no part in provision of generation ancillary services (i.e. black start, frequency control and generation reserve). Clearly it is technically impossible for intermittent renewable generators, such as wind power or photovoltaics, to increase their output above the instantaneous available energy resource. However, it is possible to arrange for their output to be reduced in response to system conditions. It is likely that this degree of control will be required on the new offshore (150 MW) wind farms now being planned in Denmark. If cost effective energy storage becomes available then both increase and reduction of power output from renewable energy plant will become possible and the renewable energy plants will begin to resemble conventional fossil fired generation in their response to network frequency.

The potential provision of black start services raises the interesting possibility of islanded sections of the network being supplied from energy storage devices, perhaps supplemented by renewable generation, during system restoration. If this mode of operation is permissible during restoration then it would appear to be appropriate at other times. The operation of islanded sections of network fed from fossil fuelled embedded generators is already being considered in Holland [15 and personal communication Mr J Hodemaekers, NV Remu].

The evidence of the National Grid Company to the House of Lords [3] indicated that once significant penetrations of intermittent renewable generation was connected to the system (e.g. more than 1500 MW of wind generation) then it would become necessary to purchase

S684/001/99 © IMechE 2000

additional reserve. This is clearly an opportunity for energy storage perhaps in conjunction with improved forecasting of the wind energy resource

4   CONCLUSIONS

There are a number of significant opportunities for increasing the value of the output of renewable generation using energy storage. A major opportunity is, of course, in Energy Trading although this would appear to apply equally to both renewable and conventional generation.

The most immediate opportunity associated with increasing the value of Transport Services is when connection of a renewable generation plant would not be permitted without major increased connection costs due to power quality or voltage rise constraints. This opportunity is enhanced by the present UK practice of deep charging for the connection of dispersed generation where the generator must pay all charges associated with the connection of the plant. A change to "shallow charging" and the application of distribution use of system charges would alter this situation and might make the use of energy storage at other locations in the system more attractive.

Further into the future, one can consider the role that renewable energy generation, with energy storage, can play to improve distribution network performance. The emphasis placed by the Regulator on Customer Minutes Lost is of particular significance as these can be reduced significantly by embedded renewable generation, supported by storage, and so the measured performance of the network may be improved. There is likely to be some benefit gained from loss reduction by reducing peak power flows but commercial gains will probably be modest.

It has been recognised that a significant amount of renewable generation will be located away from load centres and there is likely to be a concentration in the North of the UK. Thus, in the longer term, energy storage is likely to play an important role in improving the utilisation of the transmission network.

Finally, once intermittent renewable generation assumes a significant role in the power system then it will be necessary for it to play a part in generation ancillary services if the costs of operating conventional thermal plant for this purpose are to be avoided.

REFERENCES

[1]   DTI Consultation Paper, "New and renewable energy, prospects for the 21st century", Conclusions in response to the public consultation, February 2000.
[2]   Earp R, " Embedded Generation and the Transmission System", Contribution to IEE one day conference in the "Hot Topic" series, Savoy Place, 28 February 2000.
[3]   House of Lords Select Committee on the European Communities "Electricity from Renewables" HL Paper 78-I, June 1999. Memorandum of Evidence from the National Grid Company HL Paper 78-II pp 274-285.
[4]   House of Lords Select Committee on the European Communities "Electricity from Renewables" HL Paper 78-I, June 1999. Paragraph 61.

[5] CIGRE Working Group 37-23 "Impact of increasing contribution of dispersed generation on the power system", Final Report to be published in CIGRE journal ELECTRA.

[6] DTI Consultation Paper, "New and renewable energy, prospects for the 21st century", April 1999.

[7] Commission of the European Union, "Energy for the future: renewable sources of energy", White paper for a community strategy and action plant (COM(97) 599).

[8] Ruddell A. J., "Energy storage for renewable energy integration in electrical networks", Proceedings of the EESAT 98 Conference, Electrical Energy Storage Systems Applications and Technologies, 16-18 June 1998, Chester, England pp 247-252.

[9] Price A., "Why store energy when it is cheap", Proceedings of the EESAT 98 Conference, Electrical Energy Storage Systems Applications and Technologies, 16-18 June 1998, Chester, England pp 297-303.

[10] Kleimaier M., Stephanblome Th. and Hennig E., "Technical and economical feasibility of energy storage systems for utility applications" Proceedings of the EESAT 98 Conference, Electrical Energy Storage Systems Applications and Technologies, 16-18 June 1998, Chester, England pp 311-318.

[11] Dick E. P. and Maureira H. A., "Transformer station energy storage" Technical and economical feasibility of energy storage systems for utility applications" Proceedings of the EESAT 98 Conference, Electrical Energy Storage Systems Applications and Technologies, 16-18 June 1998, Chester, England pp 319-323.

[12] Baker J. N. and Collinson A. "Electrical energy storage at the turn of the Millennium", IEE Power Engineering Journal Specula Feature Electrical Energy Storage, Volume 13, No 3, June 1999.

[13] Jorgensen P., Gruelund Sorensen A., Falck Christensen J., Herager P., "Dispersed CHP units in the Danish Power System", Paper No 300-11, CIGRE Symposium "Impact of Demand Side Management, Integrated Resource Planning and Distributed Generation" held at Neptun, Rumania, 17-19 September 1997.

[14] Jenkins N., Strbac G., "Impact of Embedded Generation on Distribution System Voltage Stability", IEE Colloquium on Voltage Collapse, London, 24 April 1997, IEE Digest No 1997/101.

[15] CIRED "Preliminary report of CIRED Working Group 04, Dispersed Generation", CIRED International Conference on Electricity Distribution, June 1999.

S684/002/99

The Regenesys™ energy storage system

A C R PRICE
National Power plc, Didcot, UK

SYNOPSIS

Innogy Limited, a subsidiary company of National Power, has completed the engineering design for a 120 MWh *Regenesys*™ energy storage plant, which is to be built adjacent to a power station site in the UK. The plant will be housed in a low-rise building and, in respect of its energy storage capacity, will be the largest energy storage plant of its type in the world. The construction programme is approximately 12 months' duration and is planned to start in the spring of 2000.

The Regenesys system is a new energy storage technology, offering great flexibility in its power and energy storage rating. It has a high speed of response; supplies real and reactive power and is therefore suited to many different applications on a power system. The technology is environmentally benign, modular, comparatively easy to site and separates the power rating from the energy storage capacity. These features make it suitable for energy storage applications in the 5–500 MW range with discharge requirements from fractions of a second to 12 hours or more.

A study programme is in progress to investigate the integration of a Regenesys energy storage plant with wind generation. Subject to a satisfactory outcome, it is expected that subsequent energy storage plants will be associated with wind and other renewable generation projects

1   THE NEED FOR ENERGY STORAGE

Power systems are planned on the basis of meeting the maximum predicted demand with a reserve margin. This results in a low utilisation of some generating plant and an under utilisation of transmission and distribution plant. The operation of power systems is aimed at instantaneously balancing generation to demand plus losses. Maintaining this balance can compromise efficiency and hence the cost of electricity production.

At the simplest level, energy storage is used to balance fluctuations in the supply and demand. Over short time periods (of say less than 1 s) the requirement is essentially frequency control. Over longer time periods the requirements become those of energy management or provision of a contingency against an undesired event. Pumped hydro storage, for example, is used to balance supply and demand over periods of hours and days, and some forms of energy storage (such as those based on hydrogen) have been proposed for inter-seasonal storage.

An energy storage device can have a number of applications on a power network as shown in the following table:

Generation duties	Ancillary services	Transmission and distribution
Energy management	Frequency response	Voltage control
Load leveling	Spinning reserve	Power quality
Peak generation	Standby reserve	System reliability
Ramping/load following	Long term reserve	
Incorporation of renewables		

Use of energy storage brings the owner and operator a number of benefits including:
- increased utilisation of the most efficient/environmentally suitable generating plant
- increased use of transmission and distribution assets

hence

- cost savings in production and distribution of each MWh
- energy supply when generation is not available
- environmental benefit, through reduction in emissions and greater incorporation of renewables

2   THE USE OF RENEWABLES AND STORAGE

The use of storage on a power network, which may include renewable generation, can be of positive benefit in a number of different areas:

- As storage balances generation with demand over long time periods, greater use can be made of renewables which generate intermittently.
- System control issues arising from random generators (such as wind) can be mitigated with a storage device. This also means that the proportion of renewable generation on the network can be increased.
- Conventional generating units used to provide spinning reserve and other ancillary services can be replaced by energy storage, with associated cost and environmental savings.
- Generators which need to operate at constant load, such as some types of biomass, can be combined with energy storage to provide ramping and peaking duties on a local grid.
- A storage plant can support a local network during periods of "islanding".

S684/002/99   © IMechE 2000

- Storage can be used to operate transmission and distribution at higher load factors, avoiding the need for line enhancement.
- Storage can also be used for intertemporal arbitrage, during periods of non- generation by charging opportunistically from the grid and discharging when appropriate.

For completeness, attention is drawn to high dam hydro, which is renewable generation with in-built storage. Its output energy can be controlled and dispatched as and when required.

2.1 The location of storage devices
The optimum sites for storage plants are usually close to the end user. This maximises the system benefits. If there are economies of scale in the storage technology, then the storage device is placed further away from the end user. Usually it is only appropriate to place storage once on a network for self-evident reasons. One of the key influences on the siting of the storage device is its ownership: as the owner of the storage device controls its market influence.

In an isolated, non grid connected system such as on a remote farm or ranch, physical location is not an issue as the renewable generator and the end user are usually the same person. However on a power network, location and ownership can now be divorced. The storage device can be placed close to the generator, so that the discharge to the network can be timeshifted to meet peak demands or peak prices. Alternatively the generator could operate without storage and the storage device could be connected close to the consumers, and operated to meet their varying demands. In the latter case, the store could be owned by the renewable generator, the customer, or even by a third party.

Key issues in location are:

- Size of generation plant
- Generating profile
- Demand profile
- Power rating and utilisation of the transmission and distribution network
- Variation of cost with storage plant size

3 THE IDEAL CHARACTERISTICS OF A STORAGE DEVICE

Storage devices can be characterised by a number of parameters. Significant parameters in the consideration of storage for use with renewables are:

- Power rating: to match the power of the generator and the system
- Discharge period: to meet the demand profile
- Charge and discharge rates
- Response time and ramp rates
- Efficiency
- Cycle lifetime

- Siting requirements
- Capital and operating costs.

It is possible to produce a simple model of the power network that includes a renewable generator, such as a turbine or wind farm with a particular load. Using historical data for both wind output and energy output prices, the physical parameters of a storage device can be chosen so as to optimise the overall benefits. Typical results indicate a need for an energy storage device of a nearly similar power rating to the generator, and a discharge capability of several hours. A similar method can be adopted for a hybrid system, comprising a combination of renewables and conventional generation.

4   TYPES OF ENERGY STORAGE PLANT

Energy storage devices can be grouped into a number of categories [1]:

- Mechanical: such as flywheels, pumped hydro, compressed air energy storage
- Thermal: such as ice storage, hot water, molten salts
- Electrochemical: such as low temperature batteries, high temperature batteries, flow cells and fuel cells/hydrogen
- Direct : such as capacitors and superconducting magnetic energy storage

Pumped hydro and compressed air are usually geographically constrained. There are a number of flywheel systems in commercial development, but their use is usually directed towards power quality and uninterruptible power supplies. Superconducting devices are too expensive for large-scale use. Electrochemical devices such as batteries have been used on power networks for a number of years. They have a number of advantages in terms of their response time, ramp rate, ease of siting. However at the larger, utility scale the chosen technology must be capable of high power output and long discharge time if it is to be useful on a power network, working in conjunction with renewables. Flow batteries [2], also known as regenerative fuel cells or redox flow cells are an ideal candidate technology.

5   THE *REGENESYS* ENERGY STORAGE SYSTEM

The Regenesys™ system is based on regenerative fuel cell technology. Electrical energy is converted into chemical potential energy by 'charging' two liquid electrolyte solutions and subsequently releasing the stored energy on discharge [3]. In common with all direct current battery and fuel cell systems, a power converter (with a suitable transformer) is required to connect the system to an ac network.

A Regenesys energy storage plant consists of the main plant items shown in Figure 1.

S684/002/99   © IMechE 2000

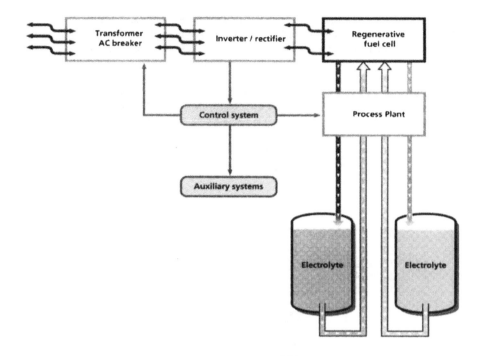

Figure 1 Block diagram of the *Regenesys* energy storage system

5.1 Regenerative fuel cell technology

Most secondary (or reversible) batteries use electrodes both as part of the electron transfer process and to store the products or reactants via electrode solid state reactions. Consequently, both energy storage capacity and the power rating are intimately related to the electrodes' size and shape. The lead acid battery is the most well known example of this type and is extensively used in many small-scale applications.

Flow cells, such as those based on zinc bromine, avoid the restricted energy capacity by storing electrolyte in separate tanks outside the cell. However the energy capacity is still limited by the mass of zinc that can physically be plated onto a carbon electrode.

Regenerative fuel cells are a separate class of electrochemical device, which have inert electrodes acting only as an electron transfer surface. The electrodes do not take part in the electrochemical process and so do not limit the energy storage capacity of the regenerative fuel cell. This approach allows the complete separation of power, determined by the module electrode area, and energy, determined by the storage tank volume.

Electrical energy is stored or released by means of a reversible electrochemical reaction between two salt solutions (the electrolytes). The Regenesys system uses electrolytes of concentrated solutions of sodium bromide and sodium polysulphide. These salts are readily soluble and present no adverse hazards in handling or storage.

The electrolytes are pumped through two separate electrolyte circuits as illustrated in Figure 2. This shows the electrolytes in the uncharged state, with sodium bromide on the positive side, and sodium polysulphide on the negative side of the cell. On charging the cell, the bromide ions are oxidised to bromine and complexed as tribromide-ions and the zero valent sulphur, present in the soluble polysulphide anion, is converted to sulphide. A cation selective membrane to prevent the sulphur anions reacting directly with the bromine separates the electrolyte solutions. Electrical balance is achieved by the transport of sodium ions across the membrane. On discharge the sulphide ion is the reducing agent and the tribromide ion is the oxidising species. The charged open circuit cell potential is about 1.5 volts, depending on the concentration of the electrochemically active species. In contrast to other systems using bromide salts, no complexing agent is required. The simplified overall chemical reaction for the cell is given by:

$$3\ NaBr + Na_2S_4 \Leftrightarrow 2\ Na_2S_2 + NaBr_3$$

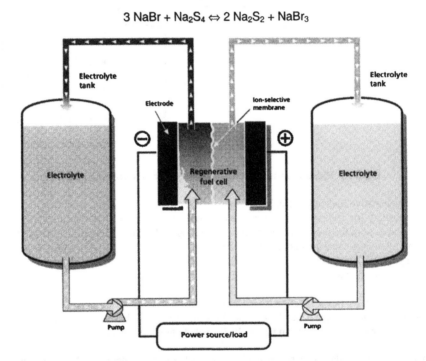

Figure 2 A single regenerative fuel cell, showing electrolyte and electrical connections

Economies of scale and manufacture can be achieved by linking cells together, with an electrode shared between two cells. The cathode of one cell becomes the anode of another. This is known as a bipolar module, and is shown in Figure 3. The module consists of parallel bipolar electrode plates, spaced and held by insulating polymer frames. These frames also serve to manifold and distribute the electrolyte into the cell compartments, which are separated by the membrane. The total open circuit voltage of a module is 1.5 x (1 +n) Volts where n is the number of bipolar electrode plates. Seals are used to prevent electrolyte leakage between cell compartments and out of the stack.

S684/002/99   © IMechE 2000

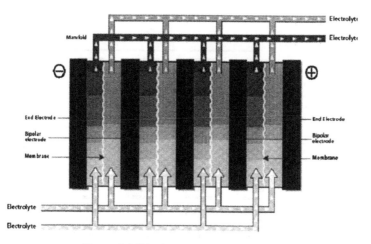

Figure 3 A Bipolar regenerative fuel cell

The conversion of electrical to stored chemical energy and back again can be repeated indefinitely with high turnaround efficiency. There is no memory effect associated with the specific electrochemistry of the Regenesys system, and a full charge/discharge cycle can be completed without limitation of a theoretical maximum depth of discharge. However, in common with all electrochemical systems, maximum efficiency is achieved below the maximum power rating for charging and a specific system can be designed to maximise efficiency once the expected operating regime has been fully characterised.

The Regenesys module, although the heart of the energy storage plant, represents only part of the complete storage system. To construct an energy storage plant a number of modules must be connected to give the required power rating. Modules are linked electrically in series to form a string of the required DC voltage and linked hydraulically in parallel. Additional strings are then added in parallel to give the required power rating of the plant (figure 4). Electrolyte storage tanks of the required volume are added to establish the energy rating of the system. The storage capacity, and hence the period of discharging electricity, is only limited by the size and number of electrolyte tanks. The storage time can be designed to suit any particular application or combination of applications. It could range from minutes to hours allowing daily charge and discharge cycles. In some circumstances such as in remote applications, discharge periods could be extended to several days.

During operation each module must be provided with a constant supply of electrolyte, recirculated from the electrolyte storage tank, through a system of distribution manifolds to the modules and back to the tanks. Electrolyte flow rates through the modules are optimised to supply the correct quantity of electrolyte to maintain the electrochemical reaction and to provide adequate cooling to the module without wasting unnecessary energy in the pumping process. The result is a system operating at low pressure and nominally ambient temperature. Ancillary systems are added to remove waste heat generated by the process and to condition the electrolytes to ensure the electrochemical efficiency remains high throughout the life of the plant. A supervisory control system ensures the integrated operation of the plant.

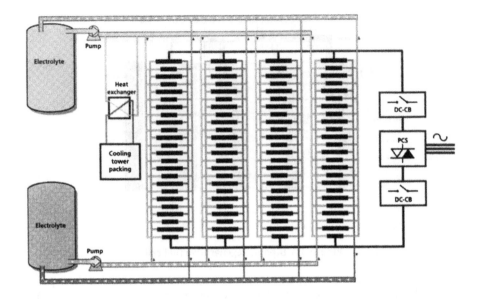

Figure 4 RFC Plant Module Array

A disadvantage of connecting cells or modules in a series arrangement is the current leakage from one cell to another and from one module to another, not through the electrodes as intended but in the electrolytes flowing through the manifolds. These shunt currents represent a loss within the system. Increasing the electrical resistance of the module manifold or pipework reduces this loss. The design of the module internal manifolds and plant layout has been developed to minimize this loss.

The principles of the technology were verified in the laboratory and later demonstrated at multi kW scale in a purpose built test facility (figure 5). The testing programme has not only verified the design, manufacturing techniques and operation of individual modules but also their ability to be connected and operated in series and parallel streams. The test facility is also used to develop and optimise the electrochemical operating regime of the system. During the development Regenesys modules of increasing size have been built and tested at the facility. The range of sizes assembled and tested is shown in figure 6.

S684/002/99   © IMechE 2000

Figure 5 The *Regenesys* operations, training and evaluation facility

Figure 6 Regenesys modules of varying sizes

The design and manufacture of the module is key to the success of the Regenesys technology. The technology has been developed over several years through a programme of work, both within and external to National Power, using consultants, universities and contractors throughout the world.

During manufacture individual bipolar electrode plates are welded at their perimeter to an electrically insulating frame. The frame incorporates much of the module detail including electrolyte distribution manifolds designed to minimise shunt currents. Several hundred bipoles with separating membranes are clamped together between solid end plates to produce a single Regenesys module. The endplates provide the connection point to the external electrolyte pipe work and electrical supplies. Individual components of the module are manufactured by specialist subcontractors to detailed proprietary designs and processes developed with National Power. Module components are delivered to National Power's assembly facility where final assembly and testing takes place. Because a large-scale energy storage plant needs several hundred bipolar modules high quality manufacturing is critically important.

Following the successful demonstration of the technology in National Power's test facility the next stage in the commercialisation of the Regenesys system is to build a full utility scale energy storage plant. The plant will be constructed alongside a power station in the UK and be rated to provide 120 MWh of energy at 14 MW power rating. A specialist contractor, Agra Birwelco, has completed the plant's detailed design to National Power's requirements. It bears more resemblance to a small chemical processing plant than to a traditional power station, as it is based on high performance polymer pipes, fittings and pumps rather than large rotating electrical machines. An artist's impression of a typical plant is shown in figure 7.

Figure 7 Artist's impression of an energy storage plant

Site erection is programmed to take approximately twelve months, commencing in the spring of 2000. Construction proceeds in parallel with module component fabrication and assembly. During construction the Regenesys modules, electrical equipment and other components will be delivered to site and installed in the module area. The electrolytes are delivered to site in a chemically stable and inert state and pumped into the storage tanks. In normal operation they remain on site for the lifetime of the plant. The plant is environmentally benign, with no emissions.

     S684/002/99 © IMechE 2000

The plant will have a high rate of dynamic response. When running, the plant will be operated fully connected to the grid, capable of turning from a state of fully charging to fully discharging or any state in between in the order of 0.02 seconds. This performance makes the plant suitable for a number of ancillary service applications such as voltage control and frequency response. In stand-by or shutdown mode there is no self-discharge of the electrolytes stored in the tanks.

A key component in any direct current battery or fuel cell system is the Power Conversion System (PCS). The PCS for the demonstration project provides the interface between the 33kV electrical supply and the variable operating voltage of the DC modules. Following a competitive tendering process, the contract for the supply of the PCS for this project has been awarded to ABB Industrial Systems.

The PCS consists of two functionally separate and autonomous converter systems, the Chopper unit (DC/DC Converter) providing the link with the variable voltage of the Regenesys modules and the DC/AC Inverter unit (a 12 pulse 3-phase 3 level DC/AC Converter). The two converter units are interconnected by a DC link with a fixed DC voltage level. The control unit provides the means to adjust incoming and outgoing voltages and currents in real time to maintain the required energy exchange between the energy storage system and the grid. The result is a four-quadrant converter system designed to transfer both reactive and real power simultaneously and independently from each other, according to the parameters set by the operator and within the capability of the PCS.

The PCS allows the operator to select from a wide range of operating modes. The normal operating mode of a storage plant will be to follow a pre-defined schedule of current/voltage/time profiles during charge and discharge including the start up and shut down of the system. The schedule will be updated on a daily basis to define the operation for the next 24 hours. In parallel with this normal operating mode, the Regenesys storage system provides the user with the following alternative control modes that can be selected to suit the rating and location of the energy storage plant and be configured to override any pre-defined schedule if required:

- Load following, plant output can be varied to match load increases or decreases;
- Voltage control mode, the store responds to fluctuations in AC system voltage providing voltage regulation under steady state and transient operating conditions ;
- Frequency regulation, the store discharges up to its rated capacity in response to a fall in system frequency or frequency rate of change ;
- Power System Stabilisation, the store provides damping of power system oscillations in the range of frequencies 0-5Hz by monitoring frequency fluctuations and controlling the store import/export;
- Constant Var., the system provides reactive power at a constant rate;
- Constant AC power, the system charges or discharges at a constant AC power
- Commercial, operated in response to signals from the system energy or capacity price.

If required a self-commutated PCS and storage system could be configured to operate as a UPS, supporting part of a distribution system or large consumer, without the requirement for rotating plant to set frequency, or to provide a Black Start capability for conventional generating plant.

6   THE FUTURE FOR ENERGY STORAGE AND RENEWABLES

The increased use of energy storage on power networks has been limited by the availability of suitable technology. Large-scale storage has usually meant pumped hydro or compressed air. Sites for constructing new pumped hydro storage are few, limited by topography and consideration of the environmental impact of such large schemes. Compressed air energy storage is also constrained by the availability of suitable caverns or porous rock media for the air storage. Both these mechanical technologies have some delay in response time and their ability to switch rapidly from charge to discharge.

Small renewable systems, typically based on PV arrays of less than 1 kW, are being extensively deployed throughout the world. In North America, Europe and parts of Asia, these systems are considered as an alternative to either primary generation or connecting to a power grid. By contrast, many parts of the world have no power at present and have a need for any form of generation to provide the most basic services such as lighting and communication. There is a requirement for renewable generation at a variety of scales, ranging from a few watts to plant sizes of 10 - 100 or more MW, The parameters of energy storage plants to best use these renewable generators will be as varied. The key parameters will be reliability and capability to charge and discharge over long durations, so as to provide power during periods of intermittent generation. Storage plants, which have, short response times (sub cycle) and high ramp rates are valuable for system operation.

In developed countries, such as in Europe and North America, where load growth is low, new renewable generation will displace existing conventional rotating plant. Conventional rotating plant has high inertia and is used to provide spinning and standby reserve, which is essential if the network is to operate within its specified frequency and voltage. The market forecast for renewable generation is significant with many ambitious targets. PV generation for example has no inertia and therefore alternative means of stabilising the system need to be introduced. Energy storage is ideally suited to fulfilling this duty.

Windpower and photovoltaics are the two fastest growing renewable technologies and present the greatest challenge in terms of grid integration. For example Windpower is growing at 35% per year, and the total worldwide market is expected to be 48 GW by 2007, from a current installed base of 9.6 GW. Much of this growth is likely to be offshore in large clusters of turbines. The market is split almost evenly between Europe and North America. Photovoltaics are the world's second fastest growing energy source. Current annual production is 150 MW growing at 20% pa.

Modular technologies such as regenerative fuel cells offer the capability of both a high power rating and a long energy storage time as well as an excellent response time so that when needed full power can be delivered in a fraction of a second. Such characteristics are important for management of electrical power requirements in an efficient manner. Operating in a fully competitive power market may involve specific dispatch requirements which may be mitigated by the use of storage. At the generation level, energy storage can be used to provide dynamic benefits such as ramping and operating and contingency reserves. Transmission companies will be able to increase the load factor of their transmission lines and other assets. Distribution companies can use energy storage to replace or defer investment in generating and other plant (such as static VAR compensators) on their electrical network and to

S684/002/99   © IMechE 2000

incorporate embedded renewable generation. Network standards may restrict the amount of embedded generation by requiring generators to shut down in the event of network faults. Connection of renewable generation with storage would allow the generator to continue generation in an island mode, subject to satisfactory re-closure procedures.

There is a significant potential market for energy storage products in the range of several hundred MW and several hours' storage down to the multi MW level that is presently unsatisfied by existing technology. Energy storage can offer increased commercial returns to renewable generators through access to capacity payments and benefit from the time value of energy, thereby accelerating the take up of renewable generation.

7 THE COMMERCIALISATION OF THE REGENESYS ENERGY STORAGE SYSTEM

The Regenesys energy storage technology has considerable commercial potential in the UK and overseas. Since the first public announcements of the technology in early 1999 there has been considerable interest from many potential customers, including renewable generators, large and small utilities, power companies and large users of power. Further projects will be developed to meet specific requirements. Ultimately a range of plant sizes will be available, each configured to maximise the benefits for the plant owners and operators.

Construction of the first full-scale energy storage plant is planned for spring 2000. Opportunities for replicating the energy storage plant at other sites are being examined, including projects incorporating intermittent renewable sources such as wind and solar. Detailed studies with a number of interested parties, within the UK and overseas, are aimed at identifying the most promising sites for development of a utility scale renewable and storage combination. Proposals have been made to various funding agencies.

Other work includes the design and development of smaller energy storage plants, (in the < 1 MW class).

8 CONCLUSIONS

Energy storage, although not a renewable energy technology in itself, is an enabling technology that will encourage and support generation from renewables. Its adoption will be in many forms, from small-scale storage co-located with small, non grid connected renewable generators, through to large scale storage with power ratings of many MW and several hours discharge connected into a power network.

The Regenesys energy storage technology, with its characteristics of separable power and energy storage rating, is well suited for use on a power networks at the utility scale. It can be incorporated at the site of renewable generation, or placed elsewhere on the network as part of a network based energy management strategy for incorporation of renewable generation.

REFERENCES

[1] Ter Gazarian, Energy Storage for Power Systems, IEE, 1994
[2] Price et al, A novel approach to utility scale energy storage, Power Engineering Journal, June 1999, IEE
[3] Vincent and Scrosati, Modern Batteries, Second Edition, 1997, pp300 et seq

S684/003/99

The costs and benefits of electrical energy storage

A COLLINSON
EA Technology Limited, Chester, UK

SYNOPSIS

There are many energy storage technologies currently available which can be used in a variety of energy storage applications. Therefore, identifying the storage technology which best matches the application requirements can be a difficult task. Each storage technology has particular characteristics which makes it suitable in some situations, but less desirable in others. It is the purpose of this paper to highlight the key features of different storage technologies (such as batteries, flywheels, SMES, capacitors, flow-cells/fuel cells) and to match the storage technologies with the most appropriate end-use applications (including the realisation of multiple-application systems) in order to achieve an optimum cost-benefit solution.

1 INTRODUCTION

There is currently considerable interest in electrical energy storage technologies, for a variety of reasons. These include changes in the worldwide utility regulatory environment, an ever increasing reliance on electricity in industry, commerce and the home, the growth of renewables, as a major new source of electricity supply, and all combined with ever more stringent environmental requirements. The rapidly accelerating rate of technological development in many of the emerging electrical energy storage systems, together with anticipated unit cost reductions, now makes their practical application look very attractive.

The widespread adoption of renewable energy sources is, in many instances, constrained by the variable and intermittent nature of their output. Their appropriate integration with storage systems will allow for greater market penetration, with associated primary energy and emissions savings. Also, the environmental impact of electricity generation is heavily influenced by the operation of older and less efficient power plant, particularly for peak lopping purposes. The appropriate integration of storage in the electricity network will reduce the need for such plant, with corresponding primary energy and emissions savings.

The market penetration achieved by electrical energy storage systems to date has been heavily constrained by their cost and the limited operational experience, resulting in high associated technical and commercial risk. However, the potential for energy storage is increasingly evident, especially for those companies operating in a competitive electricity market. Scientific advances in combination with a global shift in emphasis towards competitive utility industries has accelerated developments and there are now growing numbers of demonstration projects around the world illustrating the versatility of energy storage.

2   UTILITY APPLICATIONS OF ENERGY STORAGE

Energy storage enables the decoupling of energy supply from energy demand. This is of particular importance to the electricity industry since electricity demand is subject to substantial hourly, daily and seasonal variations. Also, electricity supply, particularly from renewable sources, is also subject to significant variability, both short term (over a few seconds) and longer term (e.g. hourly, daily, seasonal). Here, we examine the particular requirements of energy storage systems for electrical utility networks, identifying applications in the electricity industry where energy storage may be useful.

Such applications include:
- spinning reserve
- load levelling
- integration with renewables
- frequency and voltage regulation
- network stability
- deferment of new generation, transmission and distribution equipment
- enhanced quality of supply/power quality
- enhanced overall energy efficiency
- emissions and environmental benefits
- asset management
- demand-side management

Firstly, if we sub-divide the applications into generation, transmission and distribution applications, it can be said that generation applications typically have high power and energy requirements at low cost. This makes pumped hydro still hard to beat for load leveling, spinning reserve and frequency regulation (particularly in the UK). However, electrical energy storage may be an option when considering more dispersed generation such as CHP schemes and windfarms. It is unlikely that the primary function of an electrical energy storage system will be to replace or supplement generation capacity, but systems may provide this as a secondary function when installed for other reasons.

S684/003/99   © IMechE 2000

Similarly, transmission networks require high power systems, but the energy requirement for the main transmission network application (network stability) is not large (i.e. stability is currently met by using static VAR compensation based on capacitors).

By far the biggest application area for electrical energy storage in utility applications lies within the distribution network. Key applications include:

- quality of supply/power quality
- deferral of capacity/asset management
- integration with renewables/embedded generation

2.1 Quality of supply/power quality

A loss of electricity supply can cause costly interruptions in production for commercial and industrial customers. An energy storage system can increase power reliability by providing the power demand during such a supply outage. In the event of a fault, the energy storage system would automatically disconnect the customer's building from the supply network and provide all of the power required to maintain normal operations. The system would need to provide utility-grade power almost instantaneously (i.e. within one mains cycle).

The complete loss of electricity supply is thankfully quite a rare event for the vast majority of customers in the UK and so mitigation measures against such events can normally only be justified in special circumstances (i.e. a very valuable/sensitive production process or a particularly poor-performing area of supply network). However, there is growing awareness of the effect of power quality phenomena, such as voltage dips. Voltage dips occur much more frequently than complete supply outages, with many tens of events occurring per year in some areas. Commercial and industrial customers, especially those with continuous automated production processes, often operate sensitive electronic systems that cannot tolerate voltage sags. The duration of a power sag may only be a few cycles (i.e. fractions of a second) but its effects can be costly. Microcontrollers on assembly lines and computers shut down and production and data processing suffer. In some cases, the cost of a process interruption can be equivalent to a customer's total annual electricity bill.

To cope with a complete supply interruption, an energy storage system would have a typical power rating measured in megawatts and an energy rating measured in megawatt-hours for a moderately sized industrial or commercial customer. The power rating would be slightly less for a power quality application, since some power is still available from the utility supply during a voltage dip. Also, the energy storage requirement is much less in a power quality system for voltage dip mitigation, since the duration of a voltage dip is rarely more than a few seconds.

2.2 Capacity deferral/asset management

As load demand approaches the capacity of the network, utilities need to add new lines and transformers. To allow for future growth, utilities install facilities that exceed existing load demands. Therefore, utilities under-utilise expensive distribution facilities during the first few years of service. Energy storage systems can be used to help the existing assets meet the growth in demand until such time as the level of demand fully justifies an upgrade of the existing facilities.

An energy storage system to defer installation of new distribution capacity requires power measured in megawatts and sufficient storage capacity to provide rated power for typically 1 to 3 hours. In a typical distribution facility, the energy storage system would operate about 30 times per year. The system would operate most frequently during daily high-load periods that occur during seasonal peaks.

2.3 Renewable applications

Energy storage systems have several potential applications for renewable systems. For example, an energy storage system can help to deliver renewable energy at times when it is most needed. Energy is stored when the renewable energy system produces power and the energy storage system is discharged at times of peak demand, when the rate for energy is highest, which therefore gives the renewable energy a greater economic value.

Another example is the "firming up" of renewable energy sources, by increasing the predictability of future availability (generator output). This is important because electricity suppliers must be able to guarantee the amount of power they have available for their customers. Thus, energy storage makes variable renewable sources more viable, and adds to their economic value.

An energy storage system used for the integration of renewable energy sources would require megawatts of power, with storage capacity in the 2 to 4 hour range. The energy storage system would typically go through a discharge/charge cycle once or twice a day, mainly during weekday peaks/troughs.

3 STORAGE TECHNOLOGIES FOR UTILITY APPLICATIONS

The direct comparison of storage systems is difficult since different types operate using different principles. It is therefore necessary to make a comparison on the basis of common energy storage characteristics.

Assessment criteria include:
- absolute power rating
- absolute energy rating
- energy density per unit area (footprint)
- energy density per unit volume and weight
- cycle efficiency
- permissible number of charge-discharge cycles
- shelf life
- speed of response
- reliability
- unit first cost
- operational and maintenance (O&M) costs
- unit design lifetime
- end of life disposal costs
- health and safety aspects

- standards and network connection protocols
- modularity
- siting requirements
- technical and commercial risk

Many storage technologies have been considered in the context of utility-scale energy storage. These include:

- Pumped Hydro (reference)
- Batteries (including conventional and advanced technologies)
- Superconducting magnetic energy storage (SMES)
- Flywheels
- Fuel Cell/Electrolyser Systems
- Conventional Capacitors
- Supercapacitors/Ultracapacitors

Considering storage technologies and systems for electrical energy storage, four systems have emerged as true contenders for utility-scale electrical energy storage applications in the short to medium term. These systems are:

- Batteries
- SMES
- Flywheels
- Capacitors

Each technology has its own particular strengths and operational characteristics.

3.1 Battery energy storage

Battery technology includes lead-acid, advanced lead-acid and advanced battery (i.e. zinc bromine, sodium sulphur, vanadium, etc.) technologies. Conventional lead acid is the most proven technology, but has the lowest performance quotient. However, with good battery management and a well optimised operational regime, these systems have been shown to be financially viable in some applications. Not all batteries are the same and so care should be taken in the application requirement specification to ensure that the requirements placed on the battery storage media are well matched to the battery performance. Parameters involved in the battery specification include the number of anticipated discharges, depth of discharge, rate of discharge, rate of recharge, system lifetime, etc. The main drawbacks with battery systems are the low power and energy densities, which may be a problem if the available space is limited. The advanced battery technologies offer improved power and energy densities, but do not have such a proven track record. Also, advanced battery technologies are currently more expensive, but as the volume of sales increases, the costs will fall. The zinc bromine battery is a good example of an advanced battery where the manufacturing process is inherently low cost and so volume manufacture will bring the cost of these batteries down considerably in the future.

The flexibility of batteries is illustrated in the range of applications for which they have been used, from the 10MW/40MWh CHINO installation to the sub-KW/kWh UPS systems available for computer applications. For the larger battery systems, typical power-to-energy ratings result in a storage requirement from ½ hour to 4 hours. The upper limit of 4 hours is set typically by the cost of conventional generation, whilst the lower limit is set by the discharge characteristics of the battery (the faster the energy is extracted, the more batteries are required, and hence the cost increases).

A battery storage system which is generating a lot of interest since its announcement earlier this year is the "regenerative flow-cell" battery storage system. A regenerative fuel cell electrochemical system is coupled to dual liquid electrolytes contained in separate storage tanks to create an electric circuit, as shown in Figure 1. An important feature is that the power rating is governed by the size of the cell, whilst the energy rating is governed by the size of the electrolyte tanks.

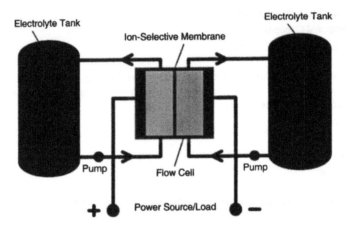

Figure 1: Regenesis regenerative flow cell

3.2 Superconducting magnetic energy storage

By comparison, SMES technology is generally still in the research and development phase, with currently one commercially available product (the American Superconductor SSD device). Other laboratory scale demonstrators are currently being evaluated for power quality applications. Superconducting technology can be neatly divided into low temperature superconductivity (LTS) and high temperature superconductivity (HTS). The main distinctions here are that superconducting wire can currently only be made out of LTS. It is easier to make coils (a key component in a SMES device) out of LTS wire than it is out of HTS ceramics. The cryogenic overheads and extra design complexities count against LTS systems, but HTS ceramics are currently very expensive. At the moment, most applications-based research uses LTS, due to the greater flexibility of the material, whilst most technology-based research is based on HTS materials to produce a cheaper material and in wire form. The ultimate goal is to achieve the implementation of the systems first developed using LTS material in HTS material, reducing both raw material and cryogenic costs.

S684/003/99 © IMechE 2000

The high cost of energy storage using SMES means that SMES is currently restricted to high-value/low-energy-requirement uses, such as quality of supply and power quality applications. Two of the main advantages of SMES is that the energy can be transferred very quickly (limited normally by the cost of the power conversion components) and unlike batteries, the number of charge/discharge cycles is virtually unlimited. SMES is thus most suitable where the storage requirement is for less than a few seconds, with power requirements up to 1 or 2 MW.

3.3 Flywheel energy storage
Flywheels are emerging as a viable technical option, sitting between batteries and SMES in terms of power/energy ratio. This makes flywheels most attractive in applications which require energy storage in the 5 second to 5 minute range. Modern high speed flywheels are made of composite materials and exhibit high rotational speeds typically in the range 20,000 to 60,000 rpm. The composite flywheel design from Urenco is for a 100kW/3kWh unit, which can be paralleled to achieve higher power and energy ratings. The feasibility of using flywheels for utility network applications looks favourable on paper, but has yet to be proven in a full-scale demonstration programme.

3.4 Capacitor energy storage
Capacitors are an option for power quality applications where the energy requirement is not large. An example of a commercially available product is the Westinghouse DVR, which is rated at 2MVA, with 400kJ of energy storage.

4   MODELLING THE COST/BENEFITS

There are many energy storage technologies currently available which can be used in a variety of energy storage applications. Therefore, identifying the storage technology which best matches the application requirements can be a difficult task. Each storage technology has particular characteristics which makes it suitable in some situations, but less desirable in others. It is important to identify the key features of different storage and to match the storage technologies with the most appropriate end-use applications (including the realisation of multiple-application systems) in order to achieve an optimum cost-benefit solution. A techno-economic cost/benefit model has been developed to aid the assessment process, based on simple spreadsheet models. The model calculates a cost-benefit ratio based on a standard "net present value" calculation.

The model allows a first-pass economic assessment of energy storage systems as applied to various utility network applications and is intended for use as a screening tool to identify those application/technology matches which merit more detailed investigation.

The storage system costs are broken down into:
- storage media cost (as a function of cost per kWh)
- power conditioning unit (as a function of cost/kW)
- balance of plant (as a function of cost/kW)

Strictly speaking, the unit costs are also a function of overall system size, since economies of scale will tend to reduce the per-unit costs. However, this is a second order effect and can be ignored in the screening model if suitable values are chosen for a typical system.

Whilst the determination of systems costs can be carried out reasonably accurately, quantifying the value of the benefits accrued is much more difficult to do accurately, since the benefits are often less tangible, separable or attributable. However, it is possible to make sufficiently accurate estimates for the purpose of identifying the most promising applications/technology matches.

The model is applications-driven, which means that a primary application is selected first. At this point, initial estimates are made for the likely system size. An energy storage technology can then be selected and the system cost is calculated based on system size and storage media. At this point, it is possible to compare the cost-benefits achievable with different storage system technologies. In addition, it is also possible to add in the additional benefits that could be accrued by so-called "secondary" benefits.

5   CONCLUSIONS

There are several application/technology matches which are worthy of consideration in utility-scale energy storage systems. Power quality applications tend to show the highest cost-benefit quotient, principally because the "energy storage" requirement is relatively low. Since the amount of stored energy required is relatively low, the power conversion and interface equipment start to become the dominant cost elements, based on power ratings rather than energy. Thus, systems for power quality applications have been realised using many different storage media (i.e. batteries, SMES, flywheels, capacitors).

Where energy storage is needed for hours rather than seconds then the choice of storage media is greatly reduced. To date, electrical energy storage systems needing multi-megawatt/mult-megawatt-hour ratings have only been implemented using lead acid batteries (except for a sodium sulphur-based systems in Japan). Energy storage systems based on lead-acid batteries have been shown to perform well in previous demonstration projects, as long as the correct type of lead-acid battery is used and good battery management is carried out. However, they are probably only commercially viable in high value applications, due to the high cost per kWh. This illustrates the importance of optimal system design, to achieve the maximum benefits from the minimum amount of necessary energy storage. However, developments in new battery technologies, especially flow cell battery technologies, which are able to decouple the system power and energy ratings by design, show great potential in applications requiring significant energy storage.

Another important concept is that of designing the system control functions in a way which can permit the realisation of "multiple benefits" from a single storage system, i.e. a system that has been configured to match several different applications concurrently. An obvious important consideration here is to ensure that the target applications do not impose mutually exclusive demands on the storage system and when conflicting demands are placed on the system that the highest priority need is met.

 S684/003/99   © IMechE 2000

ACKNOWLEDGMENTS

This work was carried out under the auspices of the International Energy Agency (IEA) within the Implementing Agreement on Energy Conservation through Energy Storage as part of the Annex IX work programme entitled *"Electrical Energy Storage Technologies for Utility Network Optimisation"*. The UK's activities within this project were co-ordinated through the Annex IX UK National Team, chaired by EA Technology.

REFERENCES

1.   **EESAT '98** *(Electrical Energy Storage Systems, Applications & Technologies)*, Chester, UK, conference proceedings, EA Technology, June 1998.

2.   **J.N. Baker & A. Collinson**, *"Electrical Energy Storage at the Turn of the Millennium"*, Power Engineering Journal, pp107–111, June 1999.

S684/004/99

The use of flywheel energy storage in electrical energy management

C D TARRANT
Urenco (Capenhurst) Limited, Chester, UK

1   INTRODUCTION

The wheel was one of mans earliest discoveries and its use to store energy is one of its most significant. Historically development advances have been governed by materials and manufacturing technology coupled with the economic need. Until recently STEEL has been the most suitable material available from which flywheels could be made with every reciprocating internal combustion engine being fitted with one. However the use of steel presents significant limitations:

*   Low Specific Strength and Low Specific Modulus leading to very large and heavy systems that run at low speed with high power losses in the bearings
*   Severe Failure Management Problems when used at high speed (Fragments as shrapnel generating casing penetration problems)

With the development in the 1970's and 1980's of fibre reinforced composite materials such as Carbon, Aramid and Glass Fibres, interest in the potential of flywheels in energy storage systems was re-kindled. It is the use of these materials in combination that gave the technology a potential way forward.

URENCO (Capenhurst) Ltd was in a unique position to develop high speed flywheels having used these types of composite materials in designing and manufacturing high speed centrifuges used for Uranium enrichment. Over the last twenty five years over 250,000 centrifuges have been constructed. This experience, combined with the invention at Capenhurst of the concept of Magnetic powder Loaded Composite materials (MLC), led to the design of a unique Flywheel' Energy Storage System (FESS). This system has potentially a large range of uses from power conditioning to energy management, this paper describes a particular application of energy management that is currently being worked on.

2   THE FLYWHEEL DESIGN CONCEPT

The development of the flywheel design commenced with an analysis of the various topologies that could be used and how these impinge on the energy storage capability, system dynamics and the failure mechanism.

2.1   Energy storage capability

The total kinetic energy of the system is given by the simple relationship $K.E. = \frac{1}{2} I \omega^2$

Where $I \propto L d^4$, $\omega_{max} \propto \sigma/\rho$. Specific strength of materials

I = polar moment of inertia; ω = angular velocity

The maximum angular velocity of the rotor, ω_{max}, is governed by the specific strength of the materials used and is the same in the hoop direction for both a cylinder and a wheel. A cylinder therefore has the potential to store more energy due to its longer length.

2.2   System dynamics

For dynamic stability the ratio of length to diameter is ideally given by:

$L/d > 3$ (a cylinder) or $L/d < 1/3$ (a wheel)

2.3   Failure mechanism

It is helpful to contrast the behaviour of a wheel to that of a cylinder when subjected to cyclic stress (ageing). When a composite material is subjected to cyclic stress, the modulus and strength of the material progressively reduce; the rate of reduction being related to the maximum stress, the ratio of the maximum to minimum stress, and the pattern of the composite lay-up.

2.3.1 Composite wheel failure

Subjecting a wheel to cyclic stress sets up the maximum stress within bore. Thus as the section is fatigued, the inner layers relax due to the drop in modulus, progressively reducing the load carried by the inner layers and thus transferring their load to the outer layers. This process eventually leads to an overload condition in the outer layers. The strength of the outer layers is reducing whilst the load they are expected to carry is increasing. This combination effect ultimately causes the wheel to fail catastrophically by bursting. This scenario has actually happened with wheels in the past and has led to some observers questioning the safety of such systems.

2.3.2 Composite cylinder

In this case the behaviour is quite different. By designing the cylinders first critical speed to be significantly below the burst speed in the event of a speed control failure, the cylinder will bend and rub on the motor generator at just below the first critical preventing the burst speed from being reached. Additionally, the reduction in modulus associated with fatigue, whilst causing the transfer of loading, also causes a reduction in critical flexural speed. The change in critical speed occurs quicker than the reduction in burst speed, hence the rotor bends allowing a system to be configured to control the cylinder's failure safely. This system is described later.

As a result of these studies, an intrinsically safe design of cylinder was chosen over the wheel. Subsequent crash tests have demonstrated the modelling to be accurate.

    S684/004/99   © IMechE 2000

3 FLYWHEEL COMPONENTS

3.1 The rotor

The rotor weighs 110 kg, is 900 mm long with a 330 mm outside diameter and a bore of 170 mm. It is constructed from three types of composite material the outer layer being carbon, the middle layer glass and the inner layer is a patented glass based Magnetic Loaded Composite (MLC). The MLC region is magnetised to form the motor/generator and a magnetic bearing.

3.2 The bearing system concept

The bearing system concept is that of a self balancing flexible rotor. This consists of a passive magnetic bearing at the top of the rotor which utilises the performance of the MLC layer and a shaft with a low loss pivot bearing mounted in a self compensating hub at the bottom. The system is design for > 10 year life without maintenance.

3.3 Containment/failure management

Containment/failure management is achieved by utilising a cylinders failure mode of bending. In the event of the rotor bending or for some other reason losing control of its centre of rotation, the operating clearances between the rotor and central shaft reduce until they eventually touch. The rotor then skids on the central shaft causing the inner layers of composite to ablate which lubricates the skidding action. The skidding action initiates a precession around the central shaft. As the precession builds up the central shaft is designed to bend causing the rotor to touch the internal bore of the outer casing. On touching the casing the rotor tries to precess in the opposite direction cancelling out/limiting the speed of precession originally set up. As a result the forces applied to the casing during a failure are limited to acceptable levels with the rotor being brought to a controlled halt, intact. This type of failure has been demonstrated on every crash experienced independent of the mechanisms used to initiate the failure.

The containment maintains the vacuum level in normal operation, but is designed to withstand the worst credible failure incident. Models of the various failure modes have been developed and a series of crash tests carried out to confirm the assumptions and predictions resulting from the modelling work.

4 ELECTRICAL POWER TRANSFER

The system is based on permanent magnet d.c. brush less technology with externally mounted solid state power switching. It acts as both a motor and generator. The power range designed and tested to date ranges from 3kW and 120kW.

The externally mounted drive system consists of a synchronised d.c. to a.c. invertor the switching of which is controlled to the rotors position by a feed back system. This position sensing system also provides speed information to a safety critical over-speed trip, which prevents the machine from being driven beyond its top speed limit. A user keypad is used to allow access to the control parameters during commissioning and subsequent data retrievals for performance monitoring. Communication with the outside world is provided by a serial communications port coupled with digital and analogue inputs/outputs which can be configured for a specific application.

The basic design concept of the system is to maintain the voltage of a d.c. bus (and hence the supply to the loads). This is achieved by the machine operating as either a motor or generator. When the machine is started the speed is increased up to a set level which is determined by the application. On reaching the speed setting the drive is turned off and the machine is allowed to coast. When due to internal losses, the speed has dropped by a pre-determined amount, the drive is turned on again to bring it back to the set point speed. Energy transfer from the flywheel to the load is triggered by the d.c. bus voltage level. If this falls below a certain level, the machine operates as a generator with only sufficient power being drawn from the machine to maintain the voltage at a constant level. The charge and discharge rates can be independently set.

5   PERFORMANCE

5.1   Cycling capability
One of the key benefits of the flywheel is its ability to handle a high number of charge/discharge cycles. The Urenco flywheel is designed to perform a at least two million charge dis-charge cycles with no effect on its performance

5.2   Speed of response
Due to the high speed of rotation and hence the high electrical switching speed the flywheel can switch from idling to either charging or discharging in less than 5 milliseconds. The actual measured values are less than 2.5 milliseconds to 95% of full power with no overshoot.

5.3   Remotely monitorable energy content
Measurement of the rotor speed is a direct measurement of the energy content.

5.4   Foot print
The relatively small foot print of the flywheel 600 mm by 600 mm by 1.5 m high.

5.5   Cost of ownership
Due to the maintenance free bearing system, low system losses hence low running cost, long potential life, and the majority of the materials being re-cycleable, the cost of ownership is low.

6   CASE STUDY – APPLICATION OF FLYWHEELS TO THE MANAGEMENT OF ENERGY IN MASS TRANSIT SYSTEMS

This study looks at the advantages that a flywheel energy storage system could provide in terms of managing the power fluctuations in a mass transit system and the potential energy savings.

6.1   Nature of the load
The service pattern of the trains is to accelerate quickly, coast for a short period and then decelerate quickly, all within a period of a few minute. This pattern generates large transient loads which are reflected back to the supply as large load swings. As at any one time there may be a large number of vehicles accelerating, load swings of 45 MW on a base load of 100 MW is not untypical with peaks as high as 75 MW. Such load swings can cause serious problems to the supply grid with excessive peakload charges and poor energy efficiency to the train operator.

S684/004/99   © IMechE 2000

6.2 Advantages of energy storage

Using flywheels coupled directly to the d.c. traction supply, the large peak power demands can be reduced. Regenerative braking energy from vehicles can be recovered, losses in conductor rails can be reduced and the system voltage regulated. These benefits improve the overall efficiency of the system reducing the cost of operation and in the case of the underground, the amount of waste heat that has to be removed from tunnels.

Additionally it has the potential to be used for emergency recovery of the trains. In the event of a supply failure and providing the system has been configured to allow it, it would be possible to use the energy stored local to the incident to provide enough energy to move vehicles out of tunnels.

6.3 System Requirements

6.3.1 Power

The following curve, figure 1, indicates power consumption for a typical train. The example used is based on an 8 car, 16 motor unit fitted with re-generative braking.

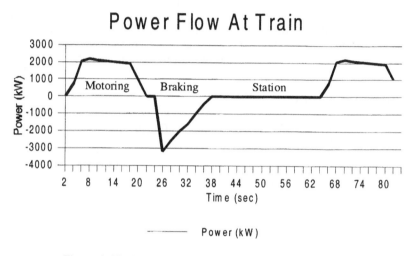

Figure 1. Train power consumption curve (seen at train)

6.3.2 System connection point

The preferred connection point would be at the stations as this would provide the energy to the load at the closest point and so minimise voltage drops and energy loss due to transmission distances. However in an underground system this may not be practical and a compromise position would have to be found.

6.3.3 Installation rating

In considering the size of system, two options are considered. First, compensation for the motoring demand only to level the load seen by the grid. Secondly, compensation for motoring demand and recovery of regenerative braking energy.

6.3.4 *Motoring*

In this case the power rating of the flywheel system is based on the mean power level, rather than the peak requirement. Thus the flywheels provide most of the accelerating power with the balance coming from the grid. The flywheels would then be recharged during the period when the train was not drawing significant power. A typical operating profile for a train is shown above. From this it can be seen that each train could present a load in excess of 2 MW for short periods with only a few 100 kW being drawn the rest of the time. A 1.2 MW flywheel system would even this out to give a continuous load at the substation. In this design of system, the flywheels are connected in parallel at the d.c. bus so that the load is shared equally between all the machines.

A block of 6 machines would have a 1.2 MW output power capability and 15 kWhr of storage. The improvement in the profile of the power requirement from the grid achieved by such a system is described in the next section.

6.3.5 *Motoring and braking*

The system described above would recover some braking energy as the flywheel would be recharging during the time when the vehicle was braking. However, because the power rating of the flywheel installation would be significantly less than the braking power of the train, not all of the energy could be recovered. With a larger system with a higher rating more of the braking energy can be recovered. It is a question of economics as to the value of the energy recovered relative to the increase cost of the larger system, as the impact on the power flow in the system would be relatively small compared to the system designed to level the demand.

6.4 Performance improvement

Approximately 60% of the available regenerative energy could be recovered by the 1.2 MW flywheel system (load levelling system). This together with the reduced system losses and reduced peak demand charge represents a significant overall saving. Figure 2 illustrates this. Doubling the size of the flywheel unit would raise the energy recovery to around 85%. However this is a relatively small saving, as illustrated by the relative areas under the curve shown, figure 2, and the economics of this have yet to be fully evaluated.

S684/004/99   © IMechE 2000

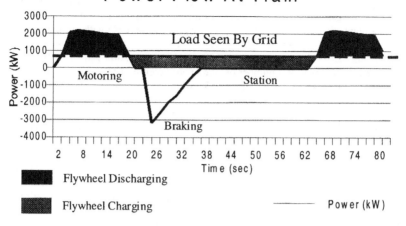

Figure 2. Resultant power flow (seen at train)

6.5 Conclusion

The connection of energy storage to the traction d.c. supply will enable significant savings in the cost of operating the system whilst at the same time compensating for the large load fluctuations currently seen in traction systems. The reduction in peak demands at the grid connection points also has the potential to allow the head way between trains on the same line to be reduced with out having to completely revamp the power supplies to the trains. This means that more trains can be run on the same system with out over loading or increasing the cost of operation. Energy storage connected to the traction d.c. supply is the best way to level the load seen by the grid. In particular, because of the large number of cycles that would be seen by any energy storage system, flywheels are the only practical solution for this application.

Whilst it is recognised that the above model is over simplified it does illustrate the potential that flywheels have in managing highly cyclic loads and or energy supplies (note: this is exactly the same reason flywheels are used in reciprocating engines). It is in this type of application that the flywheels strengths are fully utilised.

7   OTHER APPLICATIONS

As stated above Flywheels are ideal for dealing with cyclic load variations such as the case described above. However they are also being used in the power quality market where large amounts of power are required very quickly for up to 50 seconds. Typical of this type of application is in use at URENCO Capenhurst where a Flywheel system is used with a Siemens SIPCON interface to eliminate voltage dips from the supplies to one of our Uranium Enrichment plants.

7.1 Power quality

In this application URENCO is collaborating with SIEMENS in providing power quality solutions for industry. The first application of this technology is on one of URENCO's older enrichment plants that has been operating continuously since 1985 with out maintenance. The centrifuges themselves, although highly reliable, are sensitive to fluctuations in voltage level, like many other critical plants across the industrial scene. Although the effects of power quality have been known for a number of years until recently it was not thought worth trying to do something about it. However, with the deterioration seen over the last 5 years see figure, where the incidence of plant tripping has risen from an average of three times a year to an average 12 per annum. This deterioration together with the development of the flywheel at Capenhurst presented URENCO with an opportunity to build what is in effect a self financing demonstration plant.

Currently the plant is operational. The system design is of a modular construction which has allowed its progressive introduction. It consists of four 450 kW SIPCON's fitted with eight 100 kW flywheels and is designed to eliminate voltage dips down to 45% for a period of up to 28 seconds. Each SIPCON with its two flywheels protects a quarter of the plant. It can react in 2.5 milliseconds filling in the missing parts of the 50 hertz sine wave supply so that the centrifuge plant drives do not see the voltage dip.

Since the 20th May this year SIEMENS and URENCO have commenced marketing this system around the world.

8   RENEWABLE ENERGY SOURCES

One of the features of renewable energy sources are they all tend to be variable both in the short term ie seconds and in the long term ie hours. Wind, Wave, and Photovoltaic systems all suffer from these problems. This range of variability can have serious consequences for the current battery based energy storage technologies. Batteries do not like highly cyclically variable rates of charge or discharge, extremes of temperatures, etc. They can store large amounts of energy very effectively but are seriously damaged unless grossly oversized by the highly cyclic nature of the renewable energy sources. The combination of a flywheel and battery energy store in which the battery is protected by the flywheels from the short term fluctuations in both charging and discharging is an obvious area for future projects.

The question of battery protection is counter productive to the battery manufacturers as it is not in their interests to introduce a technology that will result in less battery sales.

Similarly, grid connection of the renewable technologies where the robustness of the grid smooths out the fluctuations also presents problems. By the very nature of the renewable sources, they tend to be located remotely and hence do not have robust connections to the grid. Any local supplies taken from the system will be subjected to the extremes of the system. Again, a case for energy storage to condition the supplies can be made. However it is a question of economics as to whether or not it is worth installing such a system as there are not likely to be many critical loads located adjacent to the renewable sources.

   S684/004/99   © IMechE 2000

9 SUMMARY

Flywheel Energy storage systems are not battery replacements, they are more of a complimentary technology and come into their own where the attributes of the load or supply are large and/or frequent variations. The attributes of high cyclic capability, small foot print, environmentally tolerant, low cost of ownership, fast response differentiate this technology from conventional batteries. The management of energy in systems such as the Urban Mass transit systems can have significant impact on the economics of such systems both in terms of capital cost, operating costs and the number of trains that the system can support.

10 ACKNOWLEDGEMENTS

Mr D Kelsall URENCO (Capenhurst) Ltd, Flywheel Electrical System Development Manager.
Dr D T Fullwood URENCO (Capenhurst) Ltd, Flywheel Mechanical Development Manager.
Mr A Palin URENCO (Capenhurst) Ltd, Application Manager.

Superconducting magnetic energy storage (SMES)

A M CAMPBELL
IRC in Superconductivity, Cambridge, UK

1   INTRODUCTION

It has been known for a long time that you can store energy in magnetic and electric fields. Storage using electric fields can only be short term because the finite conductivity of materials, and the low breakdown strength of insulators means that the energy density is rather low. Magnetic fields on the other hand can provide efficient energy storage for both long and short term applications and there is no magnetic breakdown strength, although there are of course limits to the magnetic fields we can generate.

The principles are simple. We build an electromagnet which is charged with magnetic field from a voltage source. When the energy is required the magnetic field collapses round the turns of the magnet generating a voltage.

If the self inductance is L and the current I the voltage is L dI/dt and the power LI dI/dt. Hence integrating over time the energy stored is $\frac{1}{2}LI^2$. This can be expressed in terms of the magnetic field as $\frac{1}{2}BH$ per unit volume, or $\frac{1}{2}B^2/\mu_o$ in free space which is the case of interest. This is independent of the detailed design of the magnet.

Associated with this energy is a pressure. If the current density in the walls is J the local Lorentz force is BJ per unit volume and $dB/dx=\mu_oJ$. The total pressure is the integral of $(B/\mu_o)dB/dx=\frac{1}{2}B^2/\mu_o$. The magnetic pressure in Pascals is numerically equal to the energy density in joules per cubic meter.

The maximum energy we can store in iron at 2 tesla with a permeability of 1000 is 1.6kJ/m^{-3} which is why we do not use iron in SMES systems. The maximum field we can generate in an air cored copper coil is about 0.1 tesla, or 4kJ m^{-3} which is also not interesting, quite apart from the continuous joule heating involved. It is only with superconductors which can produce high magnetic fields with little dissipation that that magnetic energy storage becomes a practical proposition. Although fields up to 20 tesla can be produced, such magnets consist mostly of winding and are extremely expensive. However fields of 6 tesla are easily produced by relatively simple Niobium Titanium windings and this corresponds to 14MJ m^{-3}. Although this is twenty times lower than a lead acid battery most of the energy is stored in free space and does not require constructional material. The system can be 90% efficient for a round trip of the energy, the weight low, and the power, reliability, and lifetime high. A recent review by Giese [1] gives more details and references than can be included in this paper, which has drawn heavily on this review.

2   COOLING

The main disadvantage of superconductors in energy storage is that they need to be cooled, which requires energy. This must remove not only the heat leak into the cryostat but also any losses in the superconductor. Although superconductors are resistanceless for DC currents there are hysteresis losses every time the current is changed.

The minimum energy required is set by the second law of thermodynamics and tells us that if we have to remove heat q at a temperature T and reject it to the surroundings at T_0 we need to do work $q(T_0/T-1)$. Thus one watt at 4.2K needs 70 watts cooling, at 10K, 29 watts and at 77K, 3watts. Unfortunately cryocoolers are not very efficient and real figures are about a factor of ten greater than the ideal. Typical heat leaks are of the order of a watt for large magnets so that typical coolers for 4.2K operation may need about 700 watts and at 77K about 30 watts. This is for systems on the scale of about a metre.

Although cryocoolers at present are also expensive this partly due to the relatively small market and the prices could come down by a factor of ten if a mass market developed. It is also possible to keep a magnet cold with regular deliveries of liquid gas, as is done in the frozen turkey industry, and this can be a cheaper option.

3   MATERIALS

Nearly all the superconducting magnets in the world are made of Niobium Titanium. This was discovered in the early sixties and has been continuously improved ever since. It is a ductile alloy and formed as fine twisted filaments in a copper matrix. It was developed at the Rutherford laboratory and is now used all over the world. Its critical temperature is 9K so it is used in liquid helium at 4.2K. At almost the same time Nb_3Sn was discovered with a critical temperature of 18K. This can be used at 10K, thus halving the cooling power, and has a higher critical field, but is a brittle intermetallic compound. It is therefore more difficult and expensive to produce and wind into large magnets.

Great excitement was generated in 1987 with the discovery of materials which were superconducting in liquid nitrogen at 77K. Although the excitement was certainly justified, the assumption that this would revolutionise the applications of superconductors was based on an overestimate of the difficulty of keeping magnets at 4.2K. It is much easier to keep a superconducting magnet at 4.2K than a resistive copper magnet at room temperature.

Of course if a liquid nitrogen temperature superconductor had the same properties at 77K as NbTi does at 4.2K there would be an enormous advantage, but unfortunately this has not proved to be the case. The most important parameter which enters into the calculations for an

S684/006/99

energy storage system is the critical current density of the superconductor. This can be as high as 10^6 A/cm^2 in both low and high Tc superconductors, but decreases as a power law to zero at the critical temperature and upper critical field. Thus current densities at half Tc and half the upper critical field are still high. Unfortunately the high Tc materials introduced two new features which has made them dtticult to use.

These do not include the fact that they are brittle compounds, Nb$_3$Sn has the ductility of Wedgewood China but can still be made into kilometre long wires and large magnets. The most important effect is that grain boundaries act as weak links with a greatly reduced critical current density. A grain boundary in YBCO (Ytrium Barium Copper Oxide) can carry 1000 A/cm^2 in zero field and this reduces to a negligible value in a few millitesla. This material can be used as a permanent magnet by making large single crystals, but wires for magnets are only now being developed and look as if they will be very expensive.

The other practical material is BSCCO (Bismuth Strontium Calcium Copper Oxide). For unknown reasons this material is less sensitive to the grain boundary problem in that it is only necessary to align grains in one direction to pass reasonable currents across grain boundaries. This has meant that it is possible to produce long lengths of superconducting wire on a commercial basis. This is done by packing superconducting powder into silver tubes and drawing it down to a long wire, followed by a heat treatment to sinter the grains together. In zero field this wire will carry about 20,000 A/cm^2 at 77K, but it loses the ability to carry resistanceless currents in low fields of about 0.1 tesla. To generate the sort of fields we need for SMES we need to go down to between 20 or 30K. This is considerably more economically attractive than the 4.2K needed for NbTi or even the 10K for Nb$_3$Sn, but the saving in cooling costs must be balanced against the cost of the material and the larger system needed because of the lower current density. Since the material has only recently become available it is not possible to make a sensible cost analysis at this stage except to say that all options are still open.

4   CONSTRUCTION AND OPERATION

A SMES system is just a magnet which is charged with energy by an external voltage, as illustrated in Fig. 1. If the resistance is zero the current rises linearly at constant voltage until limited by the critical current of the superconductor or the mechanical forces. If the limits are exceeded failure can be catastrophic. If a small section of the magnet goes normal the dissipation in this region spreads and causes a complete quench of the whole magnet. This can melt conductors and create kilovolts at the terminals which burn out the insulation. The bursting pressure in the bore has been set out above, but more unpredictable are the buckling forces produced by the compressive stresses at the ends. These problems have been addressed by the magnet manufacturers over the last forty years and solved. Use of magnets in SMES involves no new problems. Use of high Tc superconductors brings in new constructional problems, but these are also soluble.

Perhaps the most important point to make about superconducting magnets, like most low temperature systems, is that if they work first time they will work forever. At low temperatures things do not change with time, and the effects of cycling from room temperature should be seen in a few cycles. Although this is a new technology, if it survives a proof test at 20% above rated values on delivery it should survive the rated values indefinitely.

A simple solenoid is the most usual for SMES applications. The properties of high Tc superconductors have led to designs with a lot of small magnets rather than one large one but this is unlikely to be economic in the long term. Here the energy is stored in a region well outside the magnet itself, which is acceptable for small systems term although the stray field may have to be shielded. However if the magnet is the size of a football field we must consider the effect of this magnetic field on the environment, which will be unacceptable.

For large scale SMES systems the magnet will be in the form of a toroid which confines the field to within the toroid and avoids the buckling forces at the ends, but is a less efficient use of material.

To store energy we apply a voltage and the magnetic field builds up until we reach the maximum the magnet can stand. If the line voltage decreases then current is drawn from the magnet. The voltage is DC and decreases exponentially as the energy is drawn from the magnet. This means that the power electronics must be capable of converting a decreasing steady voltage into the line voltage at 50 or 60 Hz and not all the energy is recoverable since the voltage becomes too low.

5   SMES PROJECTS

Up to 1993 most research in SMES was on very large scale liquid helium temperature systems for diurnal storage. A number of design studies were made. There are two approaches to the engineering design. One is to use the techniques developed for smaller magnets in which the magnet is immersed in helium and the whole structure is cold. The alternative is for the structure to be at room temperature with the superconductor in tubes carrying liquid helium as the only cold part. The latter is a cheaper option but even if the magnet were to be put in a tunnel so that the rock could bear the stresses, it was concluded that the system would only be cost effective for capacities greater than 1000 MWh. Another study put the cost of a 5000 MWh system at one billion US dollars so it is not surprising that none were built.

A few years earlier Superconductivity Inc. had begun to develop a much smaller scale system to prevent voltage sags. This was a 1MJ/1MW self contained system consisting of a NbTi superconducting magnet which was mounted on a truck with all the associated cryogenics and power conversion electronics. The truck could be driven to wherever it was needed and parked in the car park. This company, now bought by American Superconductor, is a commercial provider of SMES systems. Their latest product which they call D-SMES (for distributed SMES) provides up to 20MVA, and responds within 0.5 milliseconds. The sort of application that needs this is companies like computer manufacturers where a small sag in voltage can require the rebooting of many computers over several days and is therefore very costly. American Superconductor predicts a world wide market of $500 million. The only other company with a commercial product is IGC.

The following Table 1, which is from the paper by Giese, shows the fifteen SMES projects at present being undertaken. It can be seen that nearly all use low temperature superconductors, the only high Tc ones being of small size, i.e. a few kilojoules compared with many megajoules for the low Tc materials. This is because of the long process which must take place before a new material can be used on a large scale. The BSCCO tapes which are used began as metre lengths in laboratories. New techniques needed to be developed before even short lengths could be made into magnets, and scaling up from metres to kilometres is a long and expensive process. Sufficient material to make large magnets only became available a few years ago, and the capacity to wind these wires into magnets is only now being put in place. It is therefore not surprising that there are no large scale SMES programmes using high Tc superconductors.

Whether low high Tc superconductors will be better than low Tc superconductors in SMES applications is a complex question which cannot be answered at this stage. Both are practical but the decision depends on the relative costs. Low Tc materials are much cheaper at present and carry much higher current densities, so the installation is smaller and cheaper for a given energy. However quite apart from the running costs, liquid helium coolers are much more expensive and less reliable than even 30K coolers. The costs of high Tc materials will come down rapidly with applications but they still have a long way to go before they can compete with Niobium Titanium at 4.2K.

Table 1

Country	Organisation	Material	Size	Status
Finland	Tampere University	HTS	160A/200V (5-10kJ)	Demo
Germany	Forschungszentrum, Karslruhe	LTS	300A/700Y (250)kJ)	Demo
Germany	Forschungszentrum, Karslruhe	LTS	2500A/6kV (220kJ)	First of kind
Germany	EUS GmbH	HTS 8kJ	100A/200V	Prototype
Germany	Tech. Univ. Munchen	LTS	1380A/3kV (1MJ)	Prototype
Israel	Bar-Ilan Univ.	HTS	100A/230V lkJ, 3phase	Prototype
Italy	ENEL Research	LTS	NA(4 MJ)	Prototype
Japan	Kansai Electric	LTS	350A/400V (1.2MJ)	Experimental
Japan	Kyusu Electric	LTS	1MW mod (3.6 MJ)	Prototype
Japan	ISTEC	LTS (Toroid)	20kA/2kV (360MJ)	Pilot plant
Korea	Dankook Univ.	LTS	1500A (0.5MJ)	Prototype
Korea	Korea Electrotech Inst	LTS	2kA/380V (0.7MJ)	Prototype
USA	BWX Technologies	LTS	96MW24kV (0.7MJ)	Dem
USA	Intermagnetics Gen.	LTS	750kVA (6MJ)	Com. Prod.
USA	Superconductivity Inc	LTS	2MW (3MJ)	Com. Prod.

REFERENCES

1) 'Progress Toward High-Temperature Superconducting Magnetic Energy Storage (SMES) Systems, – A Second Look'
by R.F. Giese, Argome National Laboratory, Soth Cass Avenue Argonne, Illinois 60439. (Copies from David Rose at the DTI.)

S684/006/99

S684/007/99

Hydrogen storage – technically viable and economically sensible?

D HART
Centre for Energy Policy and Technology, Imperial College, UK

INTRODUCTION

Intermittent forms of energy production, particularly renewable sources such as solar and wind power, often require energy storage to be economically and practically viable. One form of energy storage that is under consideration is the use of hydrogen – produced via electrolysis of water in many cases – which can be both economic and very clean.

1 HYDROGEN ENERGY

Hydrogen is currently used primarily in chemical processes such as the manufacture of ammonia and methanol, though a substantial amount is also used in oil refining. In the latter it is used both for hydro-cracking of heavy components into lighter ones and for hydro-treating such as the removal of sulphur. About 1% of current worldwide hydrogen production (about 500bn Nm^3 in total) is used directly for energy in space programmes. This is because hydrogen contains the most energy per unit mass of any fuel under oxidation, and minimisation of mass is a key requirement for space missions. Hydrogen contains approximately 120MJ/kg of energy, compared with natural gas at 55MJ/kg and petrol at about 42MJ/kg (approximate lower heating values).

However, the clean-burning nature of hydrogen, coupled with increasing developments in the field of fuel cells, are leading to more interest in hydrogen as a general energy carrier. Historically, as technology has developed, the trend in primary energy sources has been towards lower carbon and higher hydrogen content – from wood through coal and oil to natural gas. Extrapolating this trend suggests that non-carbon fuels could make up a significant proportion of future energy supply. Of course, renewably-generated heat and electricity will also be part of that mix, but it can be shown that they can work together with hydrogen in a synergistic system.

2   USING HYDROGEN FOR ENERGY STORAGE

Although hydrogen is found in a wide variety of compounds and is the most abundant element in the universe, it has characteristics that make it difficult to handle and store. It is a highly reactive element and is found almost exclusively in compounds such as water (H_2O) and various fossil fuels (C_xH_y), with only minuscule amounts of free hydrogen in nature. In addition it is the lightest of the elements, dissipating rapidly upwards with speeds sufficient to achieve escape velocity from the earth's atmosphere. Despite its high energy per unit mass it is also very diffuse, taking up large amounts of volume per unit of useful energy unless it is compressed, liquefied or stored in more esoteric ways. While hydrogen is flammable and explosive if mixed in the right proportions, it shares these characteristics with all other fuels. It requires some different handling and safety precautions but is not necessarily more dangerous than any other form of energy storage.

The primary advantages of using hydrogen for energy storage are its versatility in terms of what it can be produced from, and its extremely low-emissions characteristics if burned in IC engines or used in fuel cells. If burned, the only pollutant production comes from high-temperature NO_x formation from the hydrogen in the atmosphere; while use in fuel cells produces only water vapour.

3   METHODS OF HYDROGEN STORAGE

Hydrogen storage can take place in many sizes from the largest-scale salt caverns currently also used for natural gas downwards. Essentially it can be compressed, liquefied or stored in compound form, usually in a solid structure.

The very largest scale hydrogen storage is not currently used, though salt caverns are used for helium storage in the USA and have been used for town gas in the recent past (a mixture of CO and hydrogen, with up to 70% of the latter). This method of storage would be suitable for extremely large amounts of hydrogen under low pressures of a few bar, and would be comparatively difficult to recover. Some calculations suggest that up to 50% of the initial amount stored would be impossible to recover, though once the cavern was full the amounts in and out would be roughly equal.

Current large-scale hydrogen storage is carried out in compressed hydrogen tanks, usually at pressures of up to 70 bar. These are often at chemical plants or in the aerospace industry. It is also possible to use liquid hydrogen storage, which may be advantageous if large amounts are required and used at a consistent rate. If the hydrogen is not used consistently then despite the super-insulated vacuum tanks there will be some energy ingress into the liquid hydrogen. This will raise its temperature above −253°C and cause small amounts of boil-off. The vaporised hydrogen will raise the pressure in the tank and cause losses through safety venting.

Metal or liquid hydrides and carbon structures are possible alternative methods of storing hydrogen, though in most cases these are still under development. A hydride is simply a hydrogen compound such as nickel-metal hydrides used for batteries, or possibly methylcyclohexane as a liquid. The compound needs to absorb or adsorb hydrogen and then release it under certain conditions, such as when pressure is released or the substance is heated slightly. Most work in these areas is being carried out for automotive applications, though any breakthroughs will be equally applicable to stationary energy storage.

Metal hydrides typically store 1–2% hydrogen by weight, so can be heavy and may also be bulky. Some carbon structures are under investigation and show promise of storing much higher amounts, but these are strictly laboratory investigations at present. The best structures appear to be in the form of carbon nanofibres, which adsorb hydrogen by a number of mechanisms. Small samples have been tested and appear to have the capability to store 30–40% w/w, an astonishing amount. If these results can be repeated, scaled-up and commercialised then hydrogen storage on-board vehicles is a very likely possibility, and there would be tremendous spin-offs for stationary applications.

Other esoteric methods of hydrogen storage are less applicable for larger-scale energy projects. They include slush hydrogen – a mixture of liquid and solid hydrogen that is very energy dense, and such things as sponge iron and glass microspheres.

4   HYDROGEN STORAGE ECONOMICS

It is actually quite complex to consider the economics of hydrogen energy storage in general situations, as the technology is currently produced for alternative markets. The value that can be added will depend strongly on the situation.

The majority of hydrogen energy used for storage is likely to be produced from electrolysis of water using renewably generated electricity, though some much larger-scale projects may eventually be undertaken to assist in balancing load on the main electricity grid. Electrolysers are a mature technology, though considerable effort is being devoted to new development in the form of alternative materials and high-pressure systems. Coupling these with intermittent renewable power sources is already under way, with a project in Alaska using wind energy one of the first. This system will aim to take advantage of the very high prices for electricity in Alaska (US$0.50–1.00/kWh) and of the high wind resource to make what is currently an expensive system economic. In addition, most electricity in remote parts of the region is produced from diesel generators, adding a high pollution load to an otherwise pristine environment.

From this example it can be seen that the economics of a particular project will depend on many things, the capital cost of the equipment only playing a small part. In addition it is important to consider the utilisation of the equipment in terms of time spent working or idle, and the cost of alternatives. The amount of storage required will be an issue, as both the production and storage equipment may have to be oversized for peak power requirements in order to ensure that hydrogen can be produced for sufficient back-up storage. The speed of response of the equipment when required to generate electricity from the hydrogen may have an effect, as may the level of cleanliness – if NO_x is a potential problem then fuel cells may have to be used in place of IC engines.

However, there are also specific and potentially large benefits that must be included in the comparison. The cost of alternatives may be extremely high, or they may not be suited to local conditions. The use of batteries in the Alaskan winter, for example, would necessitate significant amounts of insulation or some heating to ensure efficient performance. There may be value in having the capacity to feed electricity into the grid for ancillary service uses –

voltage and frequency compensation or VAr reduction may command high prices. Equally, avoiding the extra cost of grid upgrades may make the project worthwhile, and the capacity to increase the utilisation of intermittent renewable systems should fit well with current and future greenhouse gas reduction commitments.

5   HYDROGEN SAFETY

Hydrogen has historically had a poor perception in terms of safety. It is seen as highly explosive, and although thorough scientific investigation has proved that the Hindenburg disaster was unrelated to the hydrogen used for lifting the airship, the press has been slow to take this point on board.

However, hydrogen does not necessarily exhibit any greater dangerous potential than any other fuel. It has wide flammability and detonation limits when mixed with air, but hydrogen fires burn out extremely quickly and radiate almost no heat, leaving many surrounding fixtures untouched. Although it detonates easily in enclosed spaces the detonation is likely to contain less blast energy than an equivalent petroleum or natural gas-based explosion, and in the open air hydrogen will dissipate extremely rapidly, making explosions almost impossible to generate. It is non-toxic and ingestion is not considered to be dangerous (unless the hydrogen is in liquid form, of course).

Safety standards exist for hydrogen as a chemical component but not for hydrogen in widespread energetic use. However, there is a long history of its safe use in many applications and environments, and the development of suitable standards should enable this to be maintained in energy storage. It is important to consider that the dangers it present are different, but not necessarily greater, than competing alternatives. Indeed, some forms of hydrogen storage, such as the metal hydrides, may be considerably safer.

6   THE FUTURE

The first visions of an integrated hydrogen energy system came early on in the industrial age. Jules Verne was a proponent of hydrogen energy in several of his books, and foresaw the end of the age of coal while it was at its height. However, there were also scientists and engineers who felt that the future would include hydrogen storage. In 1923 the Scottish scientist J.B.S. Haldane read a paper to a Cambridge University Society known as the Heretics, in which he said that the fuel of the future would be hydrogen. His paper described the windmills that would be used to generate electricity and the power stations where water would be split electrolytically into hydrogen and oxygen. The hydrogen would then be stored in vacuum-jacketed reservoirs before being

> *"...recombined in explosion motors working dynamos which produce electrical energy once more, or more probably in oxidation cells."*

The last comment referred to the fuel cell, which is once more under serious development. Haldane went on to address both air pollution and security of supply concerns in his vision of the future.

With the development of combined renewable-hydrogen energy systems such as the one in Alaska, it seems that Haldane's vision may be on the verge of coming true.

© 2000, with Author

Printed and bound by CPI Group (UK) Ltd, Croydon, CR0 4YY

16/04/2025

14658823-0002